Earth in Space

Rebecca Motil

Contents

Earth's Just-Right Place in Space

Earth is a special kind of **planet**—it's just right for life as we know it. The temperatures are not too hot or too cold. We have solid rock to stand on and the liquid water we need. Earth's atmosphere provides us with air to breathe and shields us from harmful radiation from space. How does Earth's place in space make it a perfect planet for us to live on?

Earth as seen from the Moon appears much bigger, and as much as fifty times brighter, than the Moon appears to us on Earth.

Earth is the third planet from the Sun. It's in what **astronomers** call the **habitable zone.** Here, the Sun's heat is just right for life. Mercury is close to the Sun and too hot for life. Liquid water would boil away on Mercury. Jupiter is far from the Sun and too cold for life. Water would freeze on Jupiter.

The Solar System

Earth is one of the planets that **orbit,** or travel around, the Sun. Together, the Sun and planets make up the **solar system.** There are also **asteroids,** which are big chunks of rock and metal, and **comets,** which are large pieces of ice, dust, and frozen gases. Of all the planets and **moons** in the solar system, only Earth has liquid water on its surface. So far, life has been found only on our special planet.

Earth's moon is also in the habitable zone. So why is there life on Earth, but not on the Moon?

It has to do with a force called **gravity.** The force of gravity pulls every object toward every other object. An object's gravity depends on its size and its mass. Because Earth is big and massive, its gravity is strong enough to hold onto the gases we need for life, such as oxygen, nitrogen, and carbon dioxide. But the Moon's weaker gravity lets these gases slip away into space.

Kick a soccer ball into the air and it will fall back to Earth because of Earth's gravity. If you kicked a soccer ball while standing on the Moon, it would travel higher and farther before falling, because the pull of gravity on the Moon is weaker than on Earth.

The Moon is a dry, airless rock in space. There is almost no atmosphere on the Moon, no liquid water, and no life.

The size of Earth is just right for holding our life-supporting atmosphere to the planet's surface. If Earth were much smaller, or less massive, the atmosphere would be too thin to support life as we know it. If Earth were much larger, or more massive, the atmosphere would be too thick.

Did you know that the Sun is a star? Life thrives on Earth because the Sun is a big star, but not too big. And it is close to five billion years old, which, for a star, is not too old. Many stars have planets, but scientists are still trying to find a planet like Earth, orbiting a star like the Sun, that can support life as we know it. When the Sun grows old, about five billion years from now, the Earth will no longer be a good place for life.

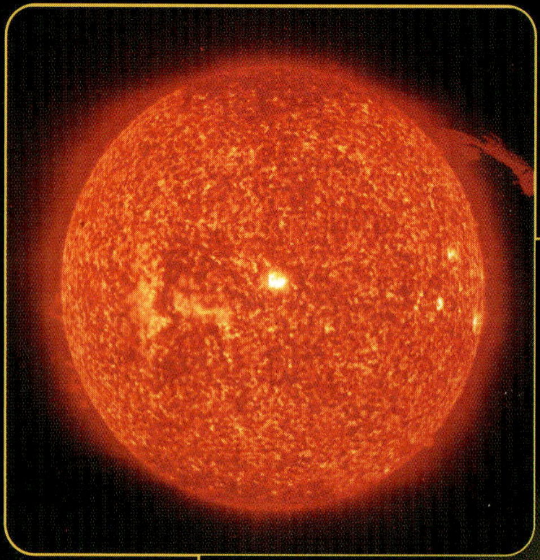

Like all stars, the Sun is a glowing mass of gas held together by its own gravity. This satellite image of the Sun shows its fiery surface, where the temperature is 10,000°F.

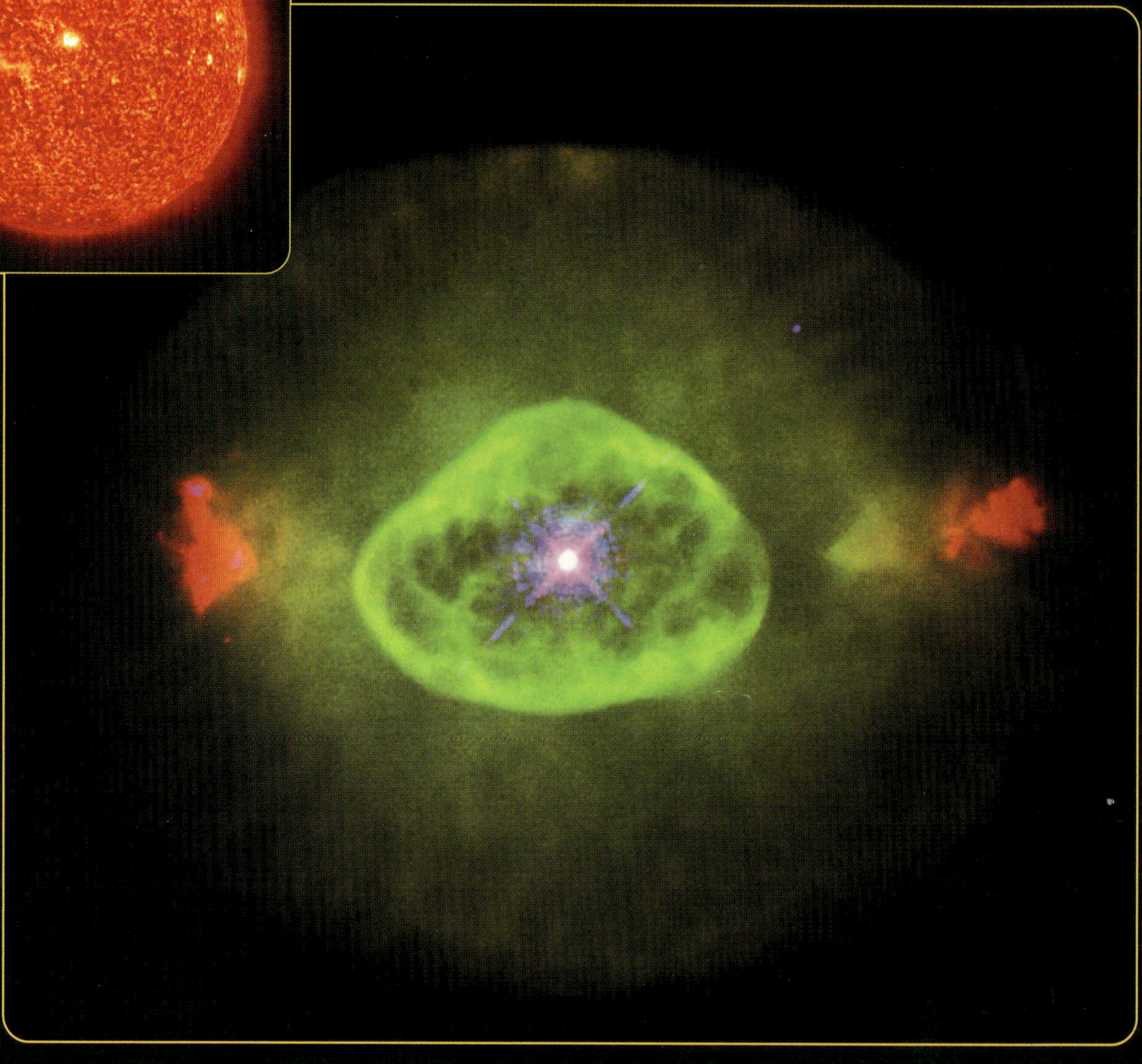

When a star like the Sun grows old, it first becomes fifty to one hundred times larger and thousands of times brighter, and then it slowly fades away. Escaping gases from a dying star can form a beautiful planetary nebula such as this one.

The Sun is one of about 200 billion stars in the Milky Way Galaxy. A **galaxy** is a collection of gas, dust, and billions of stars. The Milky Way is a spinning, spiral-shaped galaxy 100,000 **light-years** across. The Sun is in one of the arms about two-thirds of the way from the center. The arm is called the Orion Arm.

The Milky Way Galaxy is part of the **universe.** There are billions of galaxies in the universe.

Quick Fact!

If planet Earth had an address, it would be:

Earth
Third Planet from the Sun
Solar System
Orion Arm
Milky Way Galaxy
The Universe

There are no photographs of the Milky Way, but this picture shows what it looks like and about where our Sun is located. Earth's place in our galaxy is just right for life, too. In the center of the Milky Way, the stars are packed so close together that the radiation they give off would probably make life as we know it impossible. At the edge of the galaxy, there may not be enough material to make a rocky planet the size of ours.

The distant galaxies on the right were photographed using the Hubble Space Telescope. Galaxies come in different shapes. The most common are spirals, like our Milky Way, and ball- or egg-shaped ones called ellipticals.

Earthly Days

The Sun appears to rise and set once every 24 hours, giving Earth our days and nights. But it is not the Sun's motion that causes this, it is the motion of Earth. Our planet **rotates,** or spins like a top, one complete turn every 24 hours. As Earth rotates, it turns toward the east. This is why the Sun appears to rise in the east.

When our part of the globe is facing the Sun, it is daytime. When our part turns away from the Sun, night falls. Planets and other spheres that rotate turn on an imaginary line called an **axis.** The two ends of an axis are called poles.

Earth's Rotation

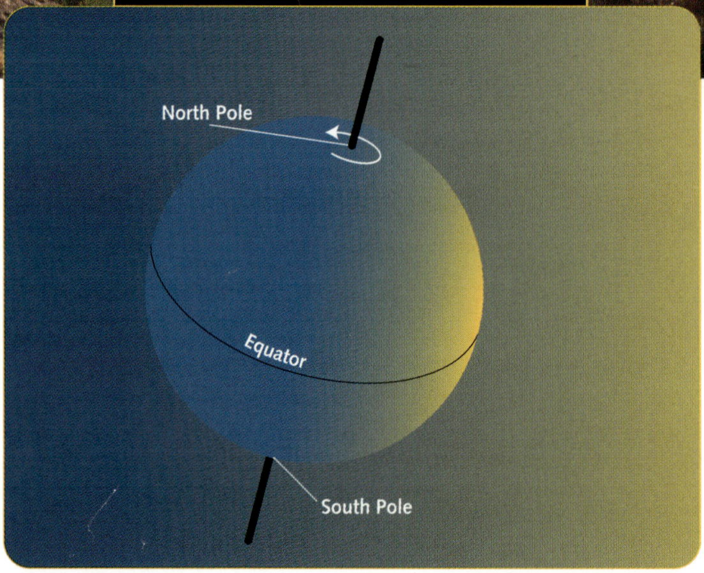

Star Trails

If you take a photograph of the night sky over a period of hours, you can capture the paths, or trails, that stars make as the Earth (and your camera on it) rotates. How the stars appear to move across the night sky depends on where you are on Earth.

Imagine standing on the North Pole. As Earth rotates, you would turn in place once every 24 hours. This would make the stars appear to move in great circles around the North Star, or Polaris, above you.

Now imagine standing near the equator, an imaginary line halfway between the poles. As Earth rotates, you would travel more than 1,000 miles per hour in a great circle about 24,900 miles around. Polaris would appear near the horizon, and the other stars would appear to rise and set, forming great arcs over it.

This is how star trails appear in California, between the equator and the North Pole. Can you find Polaris?

Earthly Years

Earth and the other planets in the solar system orbit, or travel around, the Sun. The length of time is takes a planet to complete one revolution is that planet's year. It takes Earth about 365¼ days to orbit the Sun, so our year has 365 days. What happens to the extra ¼ day? Every four years, an extra day is added to our calendar. Those years have 366 days and are called leap years.

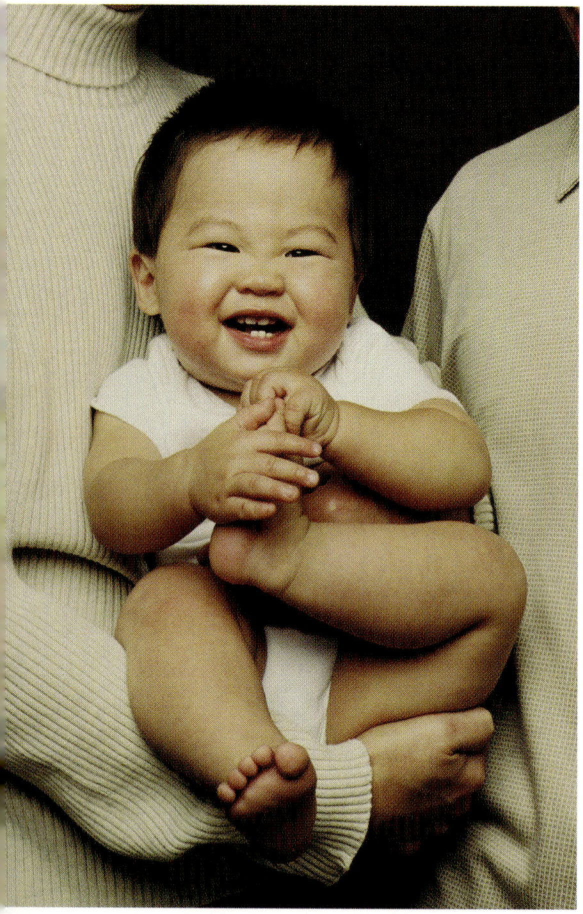

This one-year-old has orbited the Sun only once. How many times have *you* orbited the Sun?

These 14-year-olds have orbited the Sun 14 times at a speed of 66,000 miles per hour! At those speeds, why don't they fly off Earth into space? It's because they are traveling at the very same speed and direction as Earth. They are hurtling around the Sun along with Earth.

Earth's Orbit

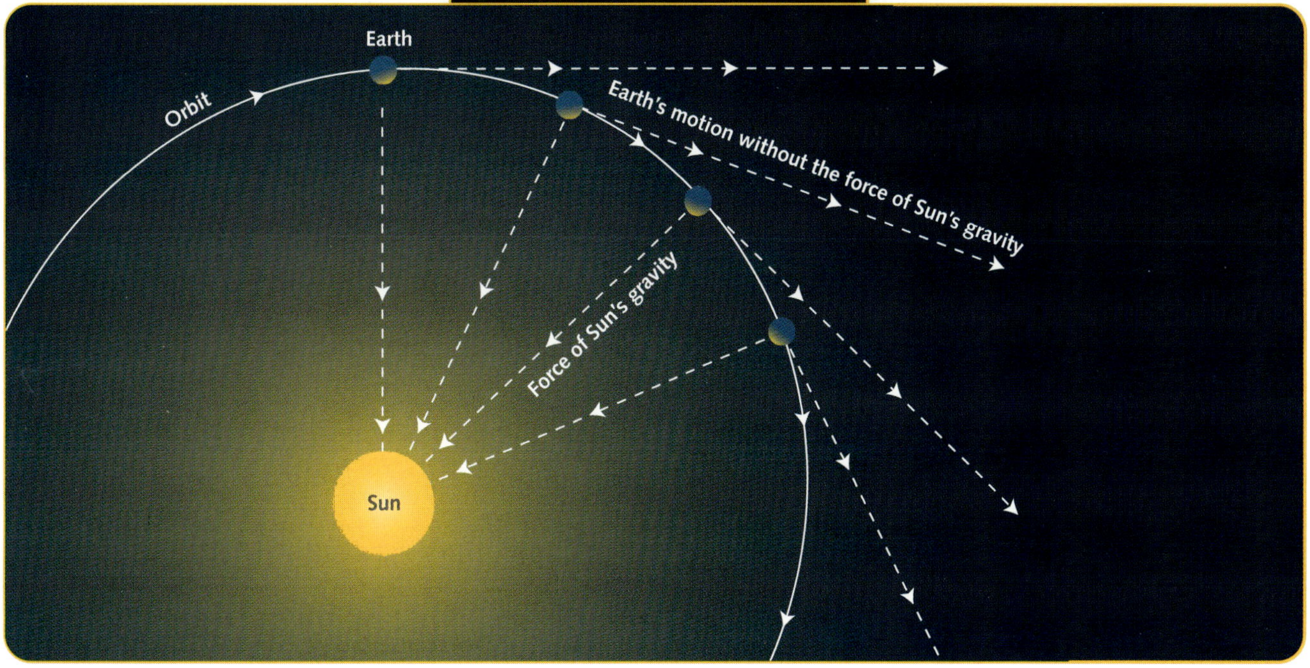

As Earth moves around the Sun, it has its own energy of motion. Without the pull of the Sun's gravity, Earth would fly off into space. The pull of the Sun bends the Earth's path into orbit around it. Earth's orbit does not form a perfect circle around the Sun; it forms an ellipse, or oval shape.

Quick Fact!

Are you standing still? If you are standing on the equator, you are moving east at almost 1,000 miles per hour as Earth rotates. You are also moving more than 18 miles per second as you (and Earth) orbit the Sun. You are orbiting the center of the Milky Way (along with the solar system) at about 125 miles per second. And you are speeding along with the rest of the galaxy at about 200 miles per second!

Earth's Moon

Earth's moon is our only natural **satellite.** A satellite is an object that is held in orbit around a larger object, such as a planet. Just as Earth orbits the Sun, the Moon orbits Earth. The Moon is held in orbit by Earth's gravity. Like Earth, the Moon spins in its orbit. However, the Moon turns once every 27⅓ days, exactly the same time that it takes to orbit Earth, so it always keeps the same side toward us.

The far side of the Moon is often called the dark side, but this is wrong. As the Moon orbits Earth, the far side gets as much sunlight as the near side. However, you can never see the far side from Earth. This photo was taken from the *Apollo 16* spacecraft.

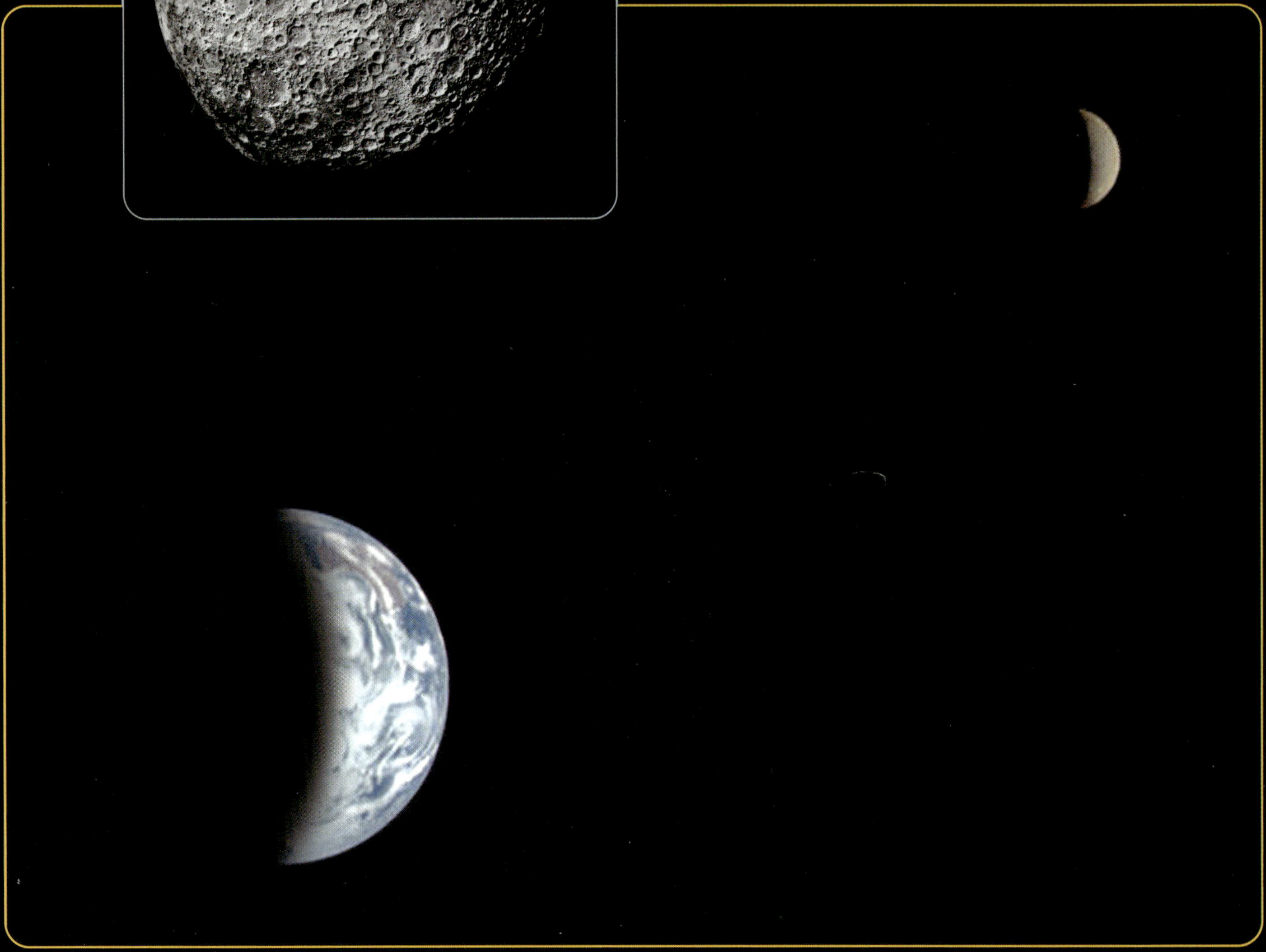

Did you know that moonlight is actually reflected sunlight? Half of the Moon is always lit by the Sun. As the Moon orbits Earth, you can see the whole lit side during the full Moon. But most of the time we see just parts of the Moon's sunlit side. The different shapes we see from the new Moon to the full Moon to a new Moon again are called **phases of the Moon.**

Quick Fact!

A thin crescent Moon is only visible close to the horizon, and the sunlit part faces the setting or rising sun. That is why thin crescent Moons are always smiley faces, never frowning ones.

Phases of the Moon

When the Moon is one quarter along its orbit, we see the first-quarter Moon in our sky. Half of the Moon's sunlit side is visible from Earth.

At new Moon, the Moon is dark in Earth's sky. This is because the sun's rays are hitting the Moon on the opposite side from Earth.

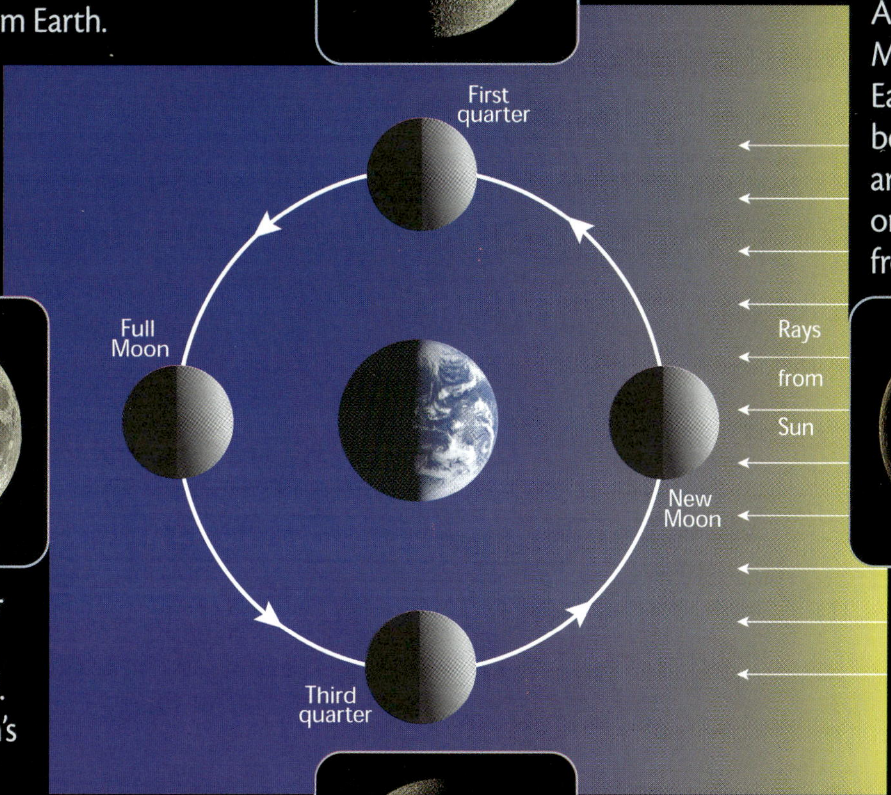

Two weeks after the new Moon, the Moon is full. All of the Moon's sunlit side is visible from Earth.

As the Moon's orbit moves another quarter turn around Earth, we see the third-quarter Moon in our sky.

What Is an Eclipse?

Imagine looking up and seeing the Sun disappear behind a dark shadow. This really happens, and it used to frighten people until they understood why. The shadow is caused by an **eclipse** of the Sun. An eclipse is the blocking of light from one body in space by another passing in front of it. From Earth we can see two kinds of eclipse: solar (Sun) and lunar (Moon).

Solar Eclipse

A solar eclipse happens when the new Moon moves directly between the Sun and Earth. The Moon's shadow has two parts, a faint outer shadow and a dark inner shadow. From the region of the faint shadow, a partial eclipse of the sun can be observed. To see a total eclipse, you must be in the dark inner shadow. Remember! Never look directly at the Sun, even during an eclipse.

In this photo of a solar eclipse, you can see the Sun's outer atmosphere, called the corona, around the edges of the dark new Moon.

A lunar eclipse occurs when Earth moves between the Sun and the Moon. During a lunar eclipse, Earth casts a large, dark shadow over all of the full Moon. That is why you can see a total eclipse of the Moon from anywhere on Earth where it is night.

During a total lunar eclipse, Earth's shadow does not blacken the full Moon. Instead, Earth's atmosphere bends and scatters some sunlight into its shadow, and the Moon becomes a reddish color.

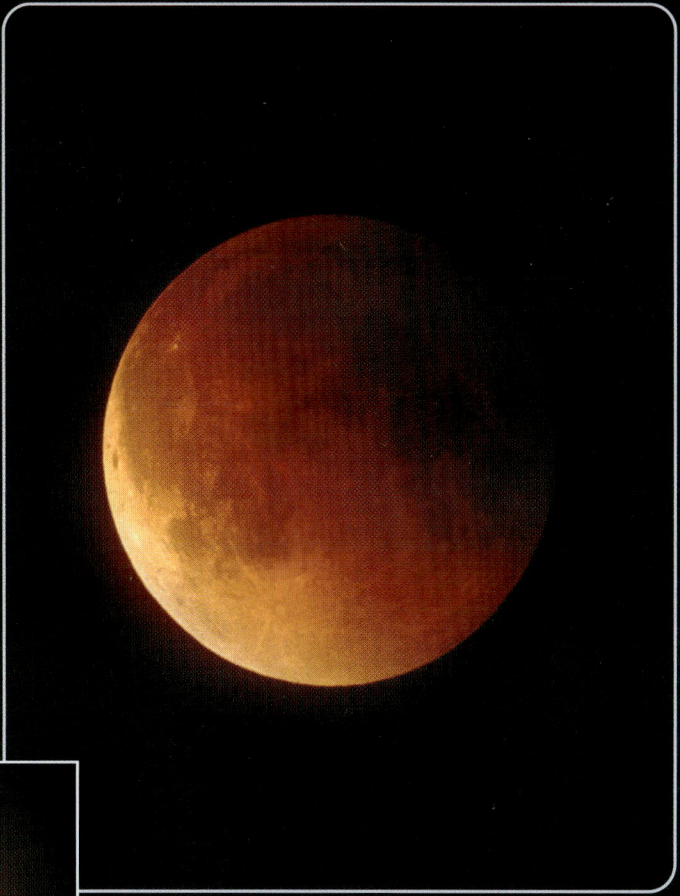

Quick Fact!

If you were standing on the Moon during a lunar eclipse, you would see a total eclipse of the Sun! It would look something like this.

Earth on Tilt

You have probably heard that Earth is tilted. But if space has no top and no bottom, how can Earth be tilted? Imagine a straight line from Earth to the Sun. That line is how light travels. If Earth's axis were straight up and down, or perpendicular to that line, our world would be a different place. But Earth's axis is tilted about 23½ degrees from the perpendicular. And that tilt gives us summers and winters.

Why Earth Has Summers and Winters

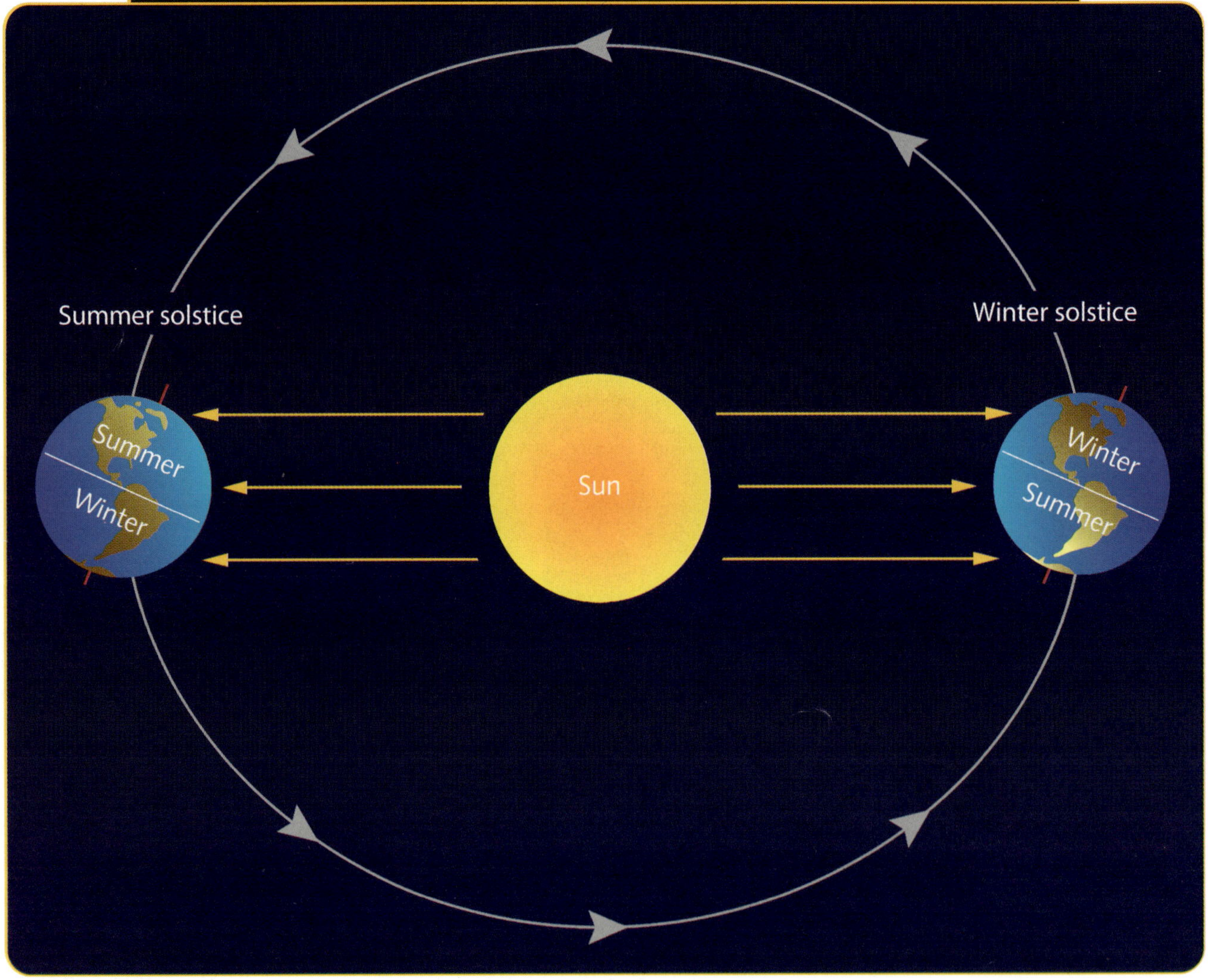

When our part of Earth tilts toward the Sun, it is summer. For us in the Northern Hemisphere, summer begins either on June 21 or June 22, our summer solstice. This day has the longest period of daylight in the year. It is winter when our part of Earth tilts away from the Sun. Our winter begins on the winter solstice, either December 21 or December 22, which has the shortest period of daylight. The seasons are just the opposite in the Southern Hemisphere.

The height of the Sun in the sky and the length of its arc change with the seasons.

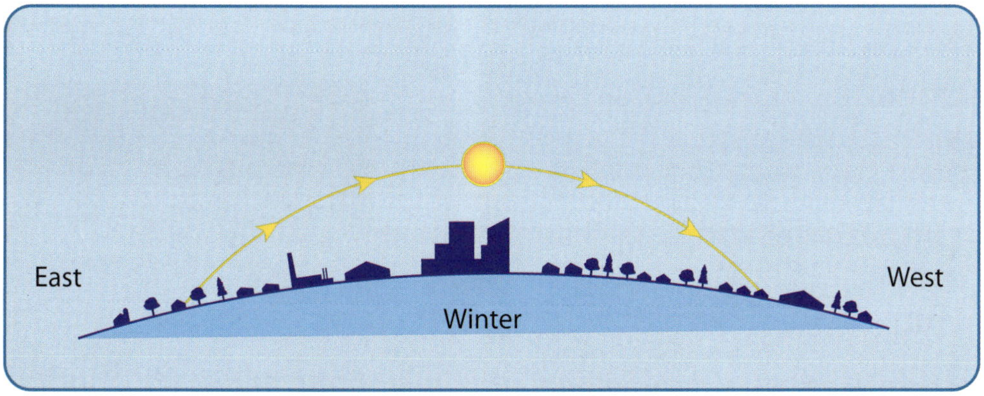

From the United States, you must face south to see the Sun's arc. Looking south, east is to your left, and west is to your right. During winter, the Sun appears low in the sky and its arc is shorter, making for shorter, cooler days.

In the summer, the Sun climbs higher in the sky and its arc is longer, causing longer, warmer days.

The Analemma

This photograph took a year to make. It captures the Sun's position at exactly the same time of day on 45 different dates throughout the year. The figure-eight shape, called an analemma, is caused by Earth's tilt on its axis, and its elliptical orbit around the Sun. If Earth were not tilted, and its orbit were a perfect circle, the Sun would reach the same point in the sky at exactly the same time every day. There would be no long days and no short days, no summer and no winter.

The amount of daylight that a place receives depends on how far north or south it is. At the North and South Poles, there are six months of continuous daylight and then six months of night.

This globe and map show daylight on December 21, our winter solstice. No sunlight reaches the far north, and in the far south, there is no darkness. In Fairbanks, Alaska, it is 6:00 A.M. and still dark.

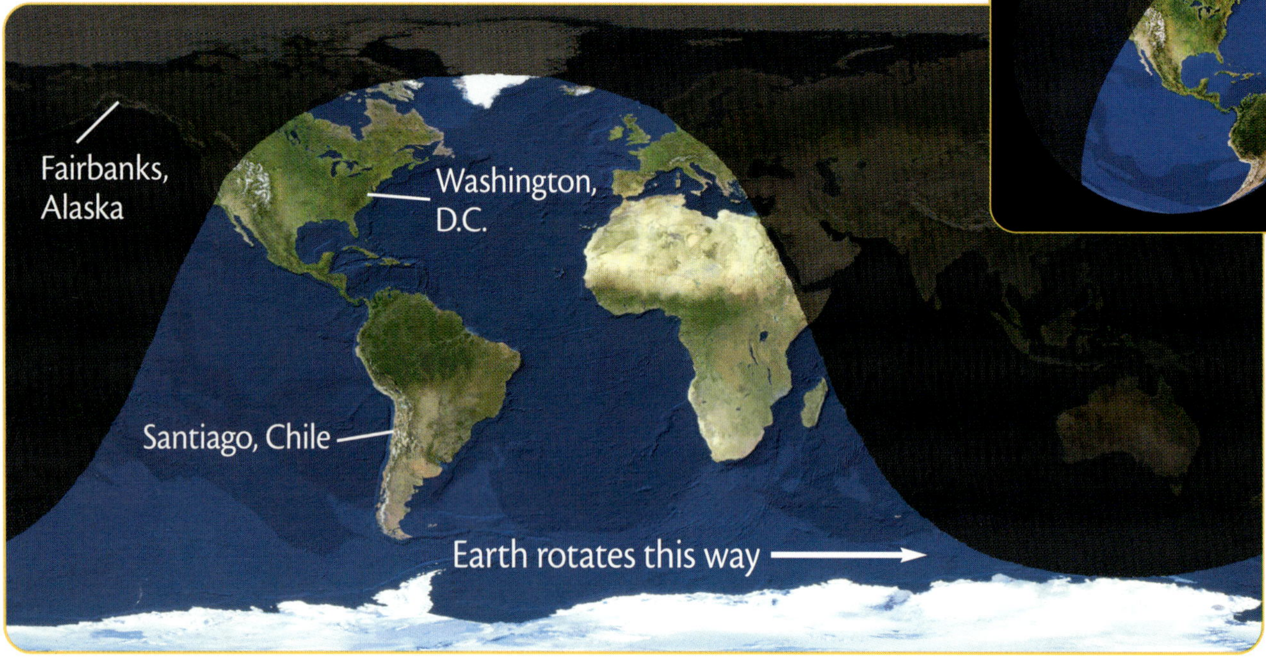

This globe and map show daylight on June 21, our summer solstice. No sunlight reaches the far south, and in the far north, there is no darkness. In Fairbanks, Alaska, it is 6:00 A.M. and the sun is shining.

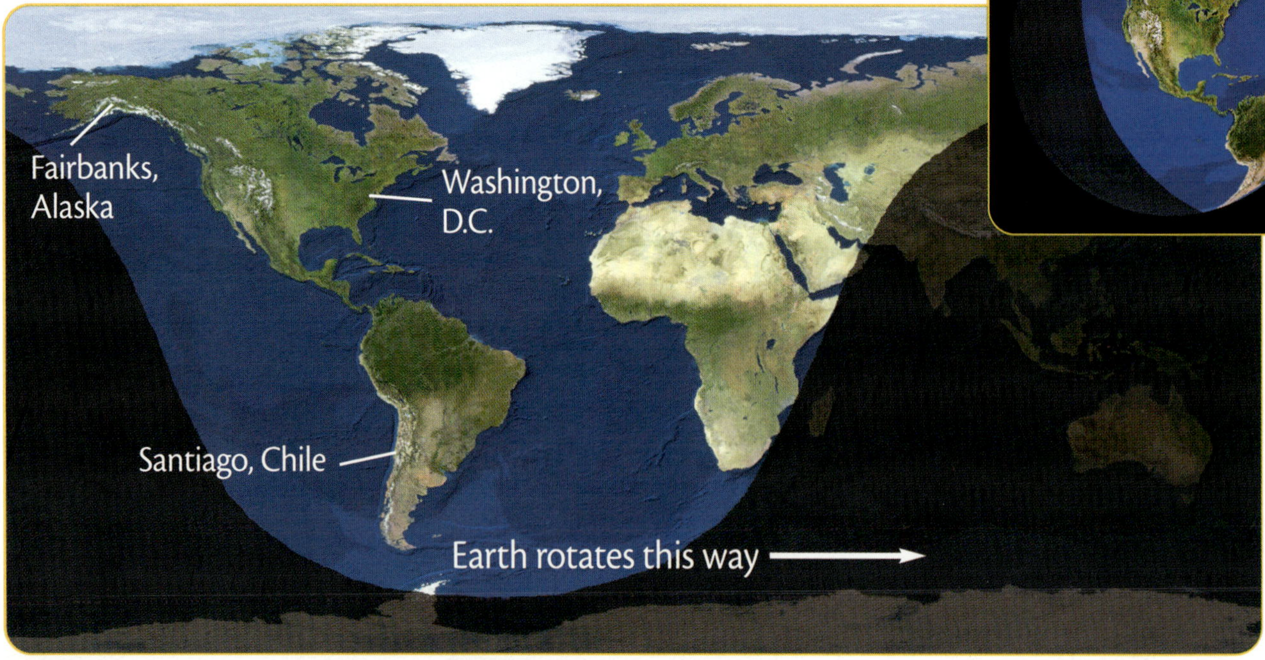

Life in the Land of the Midnight Sun

Did you ever hear of the Land of the Midnight Sun? People who live near the poles have to learn to live with periods of daylight that last for weeks or even months.

This picture was taken over a period of time in northern Alaska. It captures the Sun's position through the summer night, as it sinks toward the horizon, only to rise again.

The people of Fairbanks, Alaska, hold a Midnight Sun Run every June. The 10-kilometer race begins at 10:00 P.M., but when the runners cross the finish line it will still be light. The Sun won't set until 12:47 A.M. and, then, it will rise again at 2:57 A.M.

Glossary

asteroids (AS-tuh-roydz) chunks of rock and metal that orbit the Sun, mostly between the orbits of Mars and Jupiter

astronomer (uh-STRAH-nuh-mur) a scientist who studies the universe

axis (AK-sus) an imaginary line around which an object rotates

comet (KAH-mut) a chunk of ice, dust, and frozen gases that orbits the Sun

eclipse (ih-KLIPS) the blocking of light from one body in space by another passing in front of it

galaxy (GA-luk-see) a vast collection of gas, dust, and billions of stars

gravity (GRA-vuh-tee) the force of attraction between any objects with mass, such as the pull between Earth and the Moon

habitable zone (HA-buh-tuh-bul ZOHN) the distance around a star at which water on a planet's surface can remain liquid, and thus sustain life as we know it

light-year (LYT-yir) the distance light travels in one year, or almost six trillion (6,000,000,000,000) miles

moon (MOON) a natural satellite that orbits a planet. Our moon is the only natural satellite named the Moon.

orbit (OR-but) the path taken by one body traveling around another body

phases of the Moon (FAY-zuz uv thuh MOON) changes in the Moon's apparent shape caused by its orbit around Earth

planet (PLA-nut) a large body that does not produce its own light, and orbits a star, such as the Sun

rotate (ROH-tayt) spin on an axis

satellite (SA-tuh-lyt) an object that is held in orbit around a larger object, such as a planet

solar system (SOH-lur SIS-tum) the Sun and the planets, moons, and other bodies that orbit it

universe (YOO-nuh-vurs) all the matter, space, and time that exist